The World's Littlest Book on Climate

10 Facts in 10 Minutes about CO2

By Mike Nelson, Pieter Tans, and Michael Banks

© 2020
www.tenintenco2.com

"To our children, and our children's children, for one cool planet."

All rights reserved. No part of this publication may be transmitted or reproduced in any media or form, including electronic, mechanical, photocopy, recording, or informational storage and retrieval systems, without the express written consent of the publisher.

The World's Littlest Book on Climate: 10 Facts in 10 Minutes about CO_2
Chautauqua Institution Edition
Published by Ten in Ten CO_2
Copyright © 2020 by Mike Nelson, Pieter Tans and Michael Banks
All rights reserved.

ISBN 9798694590068

Illustrations provided by Zofostro Science and Climate Central
Cover image courtesy of NASA and needpix.com
Publication Design by Jim Primock
Website Design by Spokes Communications

Printed and bound in the United States of America

Contents

Foreword

1—CO2 on the Rise: Steadily Increasing

2—CO2 Acts as a Blanket, Trapping Heat

3—It's Us: The Global Flood of Human CO2

4—Add Heat, Planet Warms

5—CO2: The Long Goodbye

6—Your Climate Forecast: Warm Today, Hot Tomorrow

7—Honest Ice: The Great Melt

8—A Hotter Planet: More Extinction Risk

9—More Heat Fuels Wild Weather

10—CO2 Is a Human Problem with Human Solutions

About the Authors

Further Readings and Videos

Acknowledgments

Dedicated to exploring the most important issues of our times and stimulating creative responses, Chautauqua Institution is mission-bound to bring sustained focus to perhaps the most consequential issue today, climate change. Launched in 2021, the Chautauqua Climate Change Initiative seeks to advance climate education, stewardship and justice in three focal areas: our Programs, Operations and Chautauqua Lake.

Chautauqua donors have made a copy of this excellent book available as a gift to our community. *The World's Littlest Book on Climate: 10 Facts in 10 Minutes About CO2* is more than a primer. Even knowledgeable readers will learn something new about the human emissions reshaping our planet.

Learn more about the Chautauqua Climate Change Initiative at climate.chq.org

This is the world's littlest book on the world's biggest problem: CO2.

Carbon dioxide (CO2) is on the rise around the world, and it's because of us. Humans are burning fossil fuels: coal, oil, and natural gas. The global warming from these CO2 emissions has profound implications now and for future generations.

Global warming will be humankind's greatest challenge. It magnifies and multiplies the threat of every other major issue facing our society—floods, storms, harvestable land for food, inhabitable space. It can intensify existing conflicts and create new challenges in the decades ahead. Yet, addressing our climate problems could also bring out the best in humanity through solidarity, cooperation, and innovation, setting the stage for future health and prosperity for many generations to come.

1—CO2 on the Rise: Steadily Increasing

We breathe out CO2 every minute of every day. Humans and other living beings have been exhaling it for millions of years. So, what makes CO2 a problem now?

Carbon dioxide makes up a minuscule amount of Earth's air—as of 2020, a little over 412 parts per million (ppm) of CO2. When it comes to CO2, small amounts DO matter, even adding tiny amounts creates a significant impact. If there were 412 ppm of carbon **monoxide** in the atmosphere, we would be unconscious and quickly close to death. Carbon dioxide is not poisonous but it is very effective at trapping heat.

We are rapidly adding CO2 to the atmosphere. CO2 emissions are a byproduct of modern life, and that amount has been steadily increasing since the dawn of the industrial era. Everywhere, every day, every time we drive, fly, heat or cool our homes or plug in—we usually cause a little fossil fuel to be burned for energy. Burning combines the carbon in the fuel with oxygen, making CO2.

Exhibit 1
At 412 ppm, Earth's CO2 Levels Are the Highest in Thousands of Years

CO2 concentration in parts per million

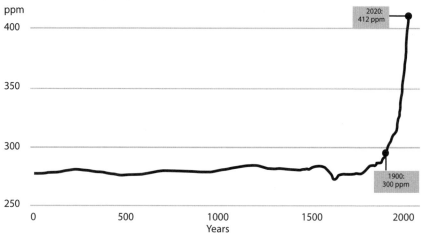

Atmospheric CO2 concentrations (ppm)
Data: CSIRO & Scripps Inst. Oceanography and NOAA Global Monitoring Laboratory
Image: Climate Central

On average, each person in the United States emits 50 times as much CO2 from burning fossil fuels as from breathing (640 pounds of CO2/person per year). This leap in the rate of emissions of CO2 is unprecedented in human history and is already greatly accelerating the changes in the Earth's atmosphere. The problem will only accelerate as our human population expands, and more people in developing countries strive for a higher standard of living and take on the living habits of developed nations.

The rate of CO2 emissions continues to increase each decade! In 1960, the world's economies emitted a total of about 10 billion tons of CO2 per year. Since then, the CO2 emissions have steadily grown and today total some 36 billion tons of CO2 per year—a more than threefold increase.

2—CO2 Acts as a Blanket, Trapping Heat

Carbon dioxide in the air works as a powerful radiant heat absorber in Earth's atmosphere—and that is good, up to a point.

The Greenhouse Effect is normal and natural. Without the heat trapping properties of CO2 and other greenhouse gases, the Earth would be about 60° F colder—a frozen ice-ball planet. The problem now is too much of a good thing—and too fast!

Unlike most of the air, CO2 is a super molecule. Oxygen and nitrogen make up most of the air we breathe, but these molecules do not trap

Exhibit 2

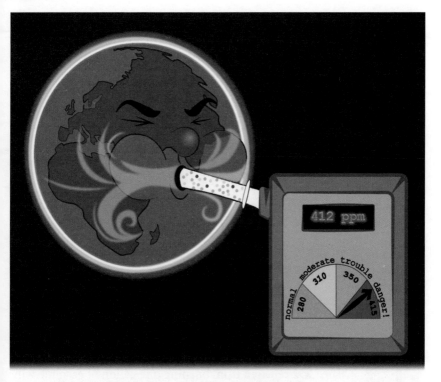

Trace levels of CO2 matter. At 412 ppm, Earth's CO2 reading is well above normal and "under the influence."

much heat. In contrast, when CO2 molecules absorb infrared light they start vibrating and rotating more strongly, but their collisions with the other air molecules (such as nitrogen and oxygen) assure that the energy of motion is shared very quickly by all. That widely shared energy is heat. The extra heat trapped in the air by each CO2 molecule causes the surface of the planet to become warmer.

Think of each molecule of CO2 as one feather in a down comforter. Each CO2 molecule added makes that blanket thicker, trapping more heat. The current increase in CO2 has a major side effect—essentially wrapping the Earth in an increasingly thick blanket and holding the heat within our atmosphere.

The heat released by the actual burning of fossil fuels is very little compared to the heat that the CO2 traps in the atmosphere. The burning of

Exhibit 3

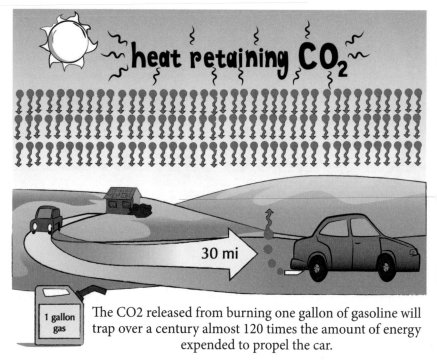

The CO2 released from burning one gallon of gasoline will trap over a century almost 120 times the amount of energy expended to propel the car.

just one gallon of gasoline creates an invisible bubble of CO2 gas almost 7 feet in diameter. That CO2 in the atmosphere will capture infra-red heat—day in and day out—for years and years. Every time we "step on the gas," the energy that speeds us down the highway is a tiny fraction of the heat energy retained in the atmosphere over time by that CO2. In 100 years, the total retained heat in the atmosphere is about 120 times larger than the combustion heat released from that original gallon of gas. By burning fossil fuels, humans unintentionally create huge amounts of atmospheric heat—over time, a giant bonfire of heat.

Current CO2 levels are not just slightly higher than normal; they are off the charts. Measurements of ancient CO2 are made by analyzing ice cores. By drilling deep into the ice sheets in Antarctica, scientists can observe "ancient air" trapped in bubbles in the ice. These tiny air bubbles have been locked in the ice for hundreds of thousands of years and enable us to know the prehistoric levels of CO2 in the atmosphere well before human emissions rose in the industrial era. These ancient ice cores show that the levels of CO2 in Earth's atmosphere averaged about 280 ppm or less. The level of CO2 has risen from about 280 ppm at the beginning of the Industrial Revolution to more than 417 ppm in 2022.

The last time that CO2 was this high in Earth's atmosphere was almost 4.1 million years ago, in the early Pliocene. Then Earth was about 7°F warmer, and ocean levels about 80 ft higher.

CO2 is not the only gas responsible for global warming. Other greenhouse gases like methane and nitrous oxide also play an important role. But CO2 currently accounts for some two-thirds of the human-caused warming. So, we cannot hope to rein in global overheating without getting a handle on CO2.

3—It's Us: The Global Flood of Human CO2

The increase in atmospheric CO2 is caused by us, humans, digging up fossil carbon and burning it for energy.

Exhibit 4
Humans Are Responsible for Today's Higher CO2 Levels
Isotopic ratios of atmospheric CO2 have been moving in the direction of fossil-fuel origins

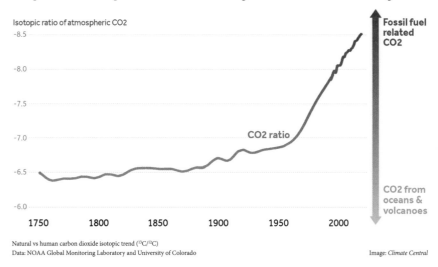

Natural vs human carbon dioxide isotopic trend ($^{13}C/^{12}C$)
Data: NOAA Global Monitoring Laboratory and University of Colorado

Image: *Climate Central*

We know that the increase in atmospheric carbon dioxide comes from humans because all CO2 has a chemical signature. From this signature, scientists can determine where it came from and how long ago it formed. Carbon exists in three slightly different forms, called isotopes. It can have an atomic weight of 12 (99% of carbon), an atomic weight of 13 (1% of carbon), or an atomic weight of 14 (radioactive, present in super-miniscule amounts).

The ratios of carbon isotopes in CO2 from fossil fuel sources are distinctly different from the isotopes in CO2 from volcanic or ocean emissions. We can accurately measure the isotopic ratios and how these have changed over time in Earth's atmospheric CO2. The CO2 emitted from fossil fuels contains a unique signature. Thus, the atmospheric measurements of carbon isotopes are the "smoking gun" proof that our burning of fossil fuels is the cause of the rapid increase of "new" CO2 in the atmosphere.

Exhibit 5
Global Temperatures Are Now Rising in Line with CO2 Levels
Higher CO2 levels lead to higher global temperatures

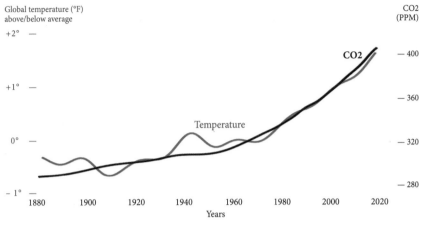

10-year smoothed global avg temperature relative to the reference period 1951–1980, & 10-yr smoothed global avg CO2 concentration
Data: NASA Goddard Institute for Space Studies (Temperature) & NOAA Global Monitoring Laboratory (CO2)   Image: *Climate Central*

4—Add Heat, Planet Warms

We understand the basic dynamics of the Earth's climate changes. Three main forces cause our planet to warm and cool, shifting between ice ages and warmer periods:

1) Varying heat from the Sun—easy to monitor and not changing much right now.
2) Changes in Earth's orbit and spin axis—well-understood but slow acting, with changes taking from 20,000 to 100,000 years.
3) Chemistry of the atmosphere—the current observed warming closely tracks the increase in global CO2 over the past 50 years.

On Earth, our emissions of extra CO2 have effectively "baked-in" a warming climate for centuries—and likely millennia—to come. Even after global CO2 emissions begin to decline, it is likely that Earth will continue heating for decades. However, it is never too late to slow this change and avoid making the warming worse.

The heat-trapping effect of the increase in CO2 is a fact. It has been known for a long time that extra CO2 in the atmosphere would warm the planet, especially in the polar regions. It is not political, and it is not new science. For example:

In 1825—Joseph Fourier, a French mathematician, realized that the Earth was much too far away from the Sun to be so warm. He thought that the atmosphere must be trapping the heat.

In 1856—Eunice Foote, an American researcher, filled different jars with different gases. She discovered that her jar with CO2 warmed the most.

In 1863—Joseph Tyndall, an Irish chemist, identified in a laboratory the heat trapping properties of CO2.

In 1896—Svante Arrhenius, the Swedish chemist, calculated that a doubling of CO2 from preindustrial levels would cause the Earth to warm by several degrees, with the high northern latitudes likely to warm twice as much.

In 1957—Charles Keeling, an American scientist, made the first precise measurements of Earth's carbon dioxide. Keeling showed the concrete evidence of rapidly increasing CO2 levels in the atmosphere. The Keeling curve is the signature reading of the steady rise of CO2 emissions from human activities.

These CO2 measurements are now taken from all over the world, but the "official" number is recorded each day from a laboratory near the top of Mauna Loa in Hawaii. The clean and well-mixed air in the middle of the Pacific provides a very precise measurement of atmospheric carbon dioxide.

The global evidence is consistent with the scientific predictions. What was once thought of as a far off problem is now at our doorstep. Adding heat from CO2 in the Earth's atmosphere is raising the world's temperature.

5—CO2: The Long Goodbye

This problem will not go away overnight. CO2 does not just fade away. CO2 is very persistent. The CO2 we are rapidly adding to the atmosphere will be here for a long time and affect generations to come. For example, about 40% of the CO2 emissions from the 1911 Ford Model T are still in the atmosphere and impacting us today.

The CO2 emissions do not disappear, but go back and forth between atmosphere, oceans, and plants. The natural processes of sequestering atmospheric carbon—rock weathering, and destruction of coral reefs, for example—work over periods of decades to hundreds of thousands of years. Humans have outpaced this natural process and that is why there is more CO2 in the atmosphere now than at any time in the recent past.

Where does all our emitted CO2 go? Currently each year we emit about 36,000,000,000 tons (or 36 gigatons [Gt]) of CO2 to the atmosphere. Because we talk generally in billions of tons, we use the abbreviation Gt as a shortcut to discuss these trends. At the end of the year there is an increase of about 18 Gt of CO2. In other words, each year about half of the emitted CO2 is being redistributed from the atmosphere to the oceans and terrestrial plants. That ratio has held fairly constant over the past five decades.

From measurements, we know that the oceans absorb about 25 percent of the world's CO2 emissions, creating increased carbonic acid that causes acidification of marine ecosystems and damages coral reefs. Another 25 percent is taken up by terrestrial plants. The dynamics of marine CO2 absorption are well known, but the impacts of terrestrial uptake are still poorly understood.

How big is the problem? Human fossil-fuel CO2 emissions are greater than the net uptake of all forests, all crops, all grasslands, all tundras, and all wetlands during the growing season in the northern hemisphere. The sheer scale of human CO2 emissions since the middle of the last century is straining Earth's natural carbon capacity. **This has all happened in the span of a single human lifetime.**

In effect, those born since World War II have enjoyed a grace period, a temporary reprieve from the climate consequences of our emissions, as it takes several decades for the temperature of the upper oceans to catch up with the increased warmth from greenhouse gases in the atmosphere. But for today's young people—and especially for those born since 2000—the grace period is going away—rapidly!

6—*Your Climate Forecast: Warm Today, Hot Tomorrow*

We can already feel our climate getting warmer—and there is more heat to come. We can think of it as three periods:

1) Since the beginning in the 20th Century, the Earth has already warmed a little less than 2° Fahrenheit from the extra CO_2 emitted into the atmosphere. In fact, about 87% of all cumulative excess CO_2 emissions have taken place between 1950 and 2020. In the past, such changes have happened over thousands of years. This change in Earth's atmosphere is taking place literally overnight in geological terms. **The years 2015 to 2020 have been the six hottest in 140 years of recorded global temperatures.**

2) The world's continued and growing CO_2 emissions also mean that the hottest years are still to come. It takes time for the extra heat to work its way through the global climate system. But the extra heat retention from the cumulative CO_2 emissions of the last several decades have "baked in" even greater warming and are projected to add another 2° to 3° F to global temperatures by 2050.

3) In addition, the engines of our global economy have a large and growing base of fossil-fuel burning cars, jets, and coal-fired power plants. The economic life of many of these machines extend 10 or 20 years, and even longer for power plants. Thus, we can expect continued CO_2 emissions at current levels into 2040 even with expected efficiency gains and replacement by renewable energy sources.

Exhibit 6
The Rate of CO2 Increase Is Rapidly Rising
The difference in CO2 concentration between the start of each decade is increasing

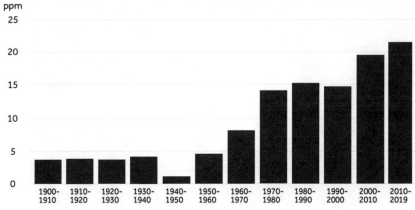

Data: NOAA Global Monitoring Laboratory Image: *Climate Central*

The cumulative buildup of greenhouse gases over the decades is forcing sustained warming of the planet, bringing our planet well above the 2.7° F of warming considered by some to be "manageable."

7—*Honest Ice: The Great Melt*

So far, the Earth's surface has only warmed a little less than 2° Fahrenheit. It does not feel all that much warmer. Sure, there are more extreme heat waves and massive forest fires, but how do we know that this is different?

Ice does not lie. Ice needs cold—freezing cold. Ice is the true measure of atmospheric cold. Across the globe, ice is in clear retreat and this shrinking ice is one of the most visible signs of our warming planet.

Exhibit 7
The North Pole: Sea ice before and after warming

Arctic sea ice at the end of summer in 1979 (above) and 2019 (below).
Photos courtesy of NASA.

Notable examples across the globe are:

- Our polar ice caps are dramatically shrinking, especially in the Arctic, as seen in the satellite photos above.

- Alpine glaciers on all continents are in steep retreat and, in fact, have begun to melt out and disappear completely.
- Winter ice cover on lakes and ponds around the world is receding and moving toward the poles.
- "Tropical" ice—the high-altitude glaciers near the equator—have clearly diminished.

The Great Melt is happening in our lifetimes. It can even be seen from space.

The greatest warming has been in the high latitudes in both hemispheres. Arctic sea ice is variable year to year, but the trend is clearly down over the past few decades. The winter sea ice in the Bering and Chukchi Seas is retreating weeks earlier than normal. Siberia experienced record heat in 2020 with temperatures reaching 100° F for the first time above the Arctic Circle, which caused early runoff that further shrank Arctic sea ice.

The great 2019 heat waves in Europe carried heat northward over Greenland. The ice loss in Greenland has nearly doubled each decade and now averages some 254 billion tons of ice loss per year. A study in 2020 concluded that the Greenland ice sheet melting has now reached the point of no return. Even a stabilization in climate warming will see its ice mass steadily decline.

The Palmer Peninsula in Antarctica has seen high temperatures shattering previous records. Recent events have shown ominous signs that its massive ice sheets are weakening and at risk of breaking loose. Some 90% of all Earth's ice is in Antarctica. Sustained melting in Greenland and the Antarctic will accelerate sea-level rise, threatening coastal cities around the world.

8—A Hotter Planet: More Extinctions

The gravest risks from a warming planet lie in the disruption to major ecosystems as well as the threats to vulnerable human populations around the world. The human cost is visible in the suffering and tragedy of climate refugees.

The consequences of this heating are already visible in the natural world around us. Increasingly, global heating poses a large and growing threat to wild species and to the integrity of ecosystems. Adding extreme heat threatens to cook ecosystems, notably in species most sensitive to temperature, such as those in the Arctic and coral reefs. Below are examples of major ecosystems beginning to fray and unravel before our eyes.

- Australia's Great Barrier Reef—Massive Coral Die Offs: A double devastation from increasing ocean temperature and from acidification. Global heating is first and foremost ocean warming. In addition, ocean surface water has already increased about 30 percent in acidity in the industrial era, with grave consequences for marine life.
- Alpine and High Mountain Ecosystems—an observed trend (also predicted by climate science) shows that earlier snow melt is causing the loss of wildflowers and longer growing seasons are resulting in the increasing dominance of woody shrubs that thrive in a warmer, dryer climate.

Ecologists are increasingly convinced that ecosystems will suffer loss and degradation as the planet heats and extreme weather events become more common. Without insects and other pollinators, well over half our food crops could not be grown. Wild plants, animals, and microbes provide the genetic library that leads us to cures for many of our illnesses and also fill our dinner tables. In a direct economic sense, we bank on a stable and temperate climate for our prosperity.

A recent Intergovernmental Science-Policy Platform on Biodiversity and Ecosystem Services (IPBES) report from the United Nations predicts that as many as 1 million plant and animal species may be at risk of extinction in our lifetimes. Our livelihoods and well-being are closely intertwined with the ecosystems that we share with other living beings.

There have been five great extinctions in the history of the Earth. Rising CO_2 levels caused or contributed to several of these. We are now seeing evidence of a sixth mass extinction around the world. Humans are changing something fundamental to the world's natural order—Earth's thermostat. Even modest changes in global temperature can have profound effects on Earth's climate—the great Ice Ages showed just a 9° Fahrenheit shift in global temperature. The actual threats from global warming to the world's ecosystems may appear subtle but could well be devastating because it is not just about wildlife. A contemporary mass extinction would have a devastating and wide-ranging impact on our own well-being.

9—More Heat Fuels Wild Weather

One degree of warming does not seem like a lot. In fact, we can barely feel a 1° difference in the temperature on any given day. But the Earth is not so simple. Climate is a complicated system and even small changes in the Earth's climate—distributing energy from the sun among our atmosphere, hydrosphere, and biosphere—can lead to major changes in our weather.

Global warming is amping up our weather systems. A warmer world evaporates more water into the air and adds more energy to super-charge tropical storms, flooding rains, and hurricanes. In times of drought, the warmer weather will make the dryness even worse. Drought is not just a lack of precipitation, it is a combination of lack of rainfall and increased evaporation due to hotter weather. The excess heat from cumulative human emissions is changing earth's weather.

How climate change translates into weather is complex. Even one or two degrees of warming is enough to trigger significant and dramatic shifts in local weather patterns. Climate change is thought to be affecting the strength and position of the jet stream with major ramifications on local weather patterns. For example, the deadly 2019 summer heat waves in Europe were connected to a shift in the North Atlantic jet stream. The record temperatures in Paris and elsewhere during that heat wave were set not by small incremental amounts but by jumps of 3, 5, or even 7° Fahrenheit.

Anchorage, Alaska experienced its first-ever 90° day on July 5, 2019 and had an unprecedented consecutive 34 days of above average temperatures. Hurricane Dorian demolished areas of the Bahamas that had survived decades of other Caribbean storms. Hurricanes on steroids such as Sandy, Harvey, Michael and Laura are high-intensity weather events that might normally occur once in a hundred years—but to have so many occur in such a short period of time is a sign that we need to plan for a new weather future.

But what does global warming mean for the average person? Ask the citizens of Houston or the wildfire victims of Colorado, California, Australia, or the hurricane-devastated Bahamians. The western U.S. wildfires of 2020 signaled a scale change in destruction. There is a climate component in these extreme-weather events that can impact each of us. The modern plagues of fire, flood, and drought are becoming more common and more destructive. Increased extreme-weather events are causing tens of billions of dollars in property damage. There is already a price to be paid for our CO_2.

The primary victims of a hotter planet will be those born today—our children and grandchildren. Many of them will live to the year 2100 and feel the full brunt of CO_2's climate impact. The recent extreme weather events coming from the added warming so far are only a sneak preview. They will live in a world where global temperatures increase by an additional 4° to 6° Fahrenheit, well beyond what the world has already experienced. Future generations will look back in disbelief that we continued to pour CO2 into the air without restrictions or restraint.

10—*CO2 Is a Human Problem with Human Solutions*

The good news is that this climate emergency is a human problem and it has human solutions. In other words, we do not have to accept this fate sitting down. We made these conditions and going forward, we have the power to change them.

Life before fossil fuels was cold, dark, and short. We owe many of the benefits of our wonderful modern way of life to fossil fuels. However the fact is that the CO2 from the burning of fossil fuels poses a grave threat to our well-being. Unless we take decisive steps to embrace different, more climate-friendly paths, our social and economic well-being will be at risk.

An effective climate solution will require a pragmatic, "all-of-the-above" approach to yield real CO2 reductions. No single energy source offers a silver bullet to effectively decarbonize our economy. Low-cost wind and solar power promise a green energy revolution everywhere, including the developing countries. But even natural gas, carbon-capture technologies, and nuclear power will have an important role to play as a bridge from coal to balance the grid. It will even require technologies that do not yet exist. We are past the point of half-hearted or perfect solutions. There is an urgency in the climate crisis created in part by our own repeated delays.

Fortunately, the shift to decarbonize our economy is already under way and gaining momentum. In 2020, almost 80% of all new power generation installed was renewable energy. We already have witnessed a rapid decrease in the cost of renewable energy. Coal is being replaced by wind and solar—not just because it is good for the environment, but because coal is now less cost effective. California, the world's 5th largest economy—is rapidly moving toward 100% renewable energy by 2045. Texas, the home of oil & gas, is already the largest wind-power producer in the US. This transition is being driven by the private sector that sees a profitable future in renewables.

The transportation industry is undergoing a dramatic change in technology as the Internal Combustion Engine (ICE) is being replaced by electric vehicles. Continuing breakthroughs in battery storage and capacity will revolutionize the industry. Electric vehicles have 90% fewer moving parts compared to ICE vehicles. It is now possible to power your car off the sunshine hitting your roof. Major companies, like GM and Ford, recognize that the automobile business is on the verge of one of the greatest revolutions in the last hundred years. Electric cars are faster, require little maintenance, and fun to drive.

Exhibit 8

We have done this before—great and transformational projects have been a part of our collective human legacy! Taking on big challenges is in our DNA. Never underestimate the innovative power of human ingenuity. The climate crisis is the great challenge of our generation. Working together to minimize climate change and restore ecosystems will bring out the best in us all, drive innovation and set the stage for future prosperity for many generations to come.

Previous generations accomplished technological advances, such as:

- The Transcontinental Railroad
- Modern Water Treatment and Indoor Plumbing
- Municipal Electrification and Rural Electrification
- The Interstate Highway System
- Apollo Moon Mission
- Creating the World Wide Web and Modern Internet

None of these projects bankrupted the United States and each of them transformed, in only a few decades, the country—and in some cases the world. Just like every other generation before ours, we need to improve and upgrade our infrastructure. By creating an efficient means of quickly moving electricity to where it is needed, we can power our nation with renewables. The revitalization of our energy systems is already taking place and can be greatly accelerated even with existing technologies. "Greening the grid" will create new economic vitality, a wave of job creation and new business opportunities.

Fifty years ago, the United States achieved the historic accomplishment of landing a man on the moon. Michael Collins, one of the Apollo 11 astronauts, said that on that historic mission what struck him most looking at the Earth floating in space was its "fragility." The sense that our planet is our space capsule. Today, over 50 years later, it is clear: climate change from increasing CO_2 is the single greatest threat to our shared and fragile planet.

* * *

About the Authors

Mike Nelson is an award-winning meteorologist in Denver, Colorado, and the recipient of 20 Emmy awards for Weather Excellence and the Edward R. Murrow Award. He is a two-time winner of the Colorado Broadcaster of the Year Award. In 2018, Mike was named a Fellow of the American Meteorological Society (AMS). Mike is the only television meteorologist in Colorado ever named an AMS Fellow and one of just 30 TV weather-casters nationwide. Mike has authored two books on weather and climate, *The Colorado Weather Almanac* (2007) and *The Colorado Weather Book* (1999). Mike now spends considerable time writing and teaching about the science of and the solutions for Global Warming. He is a graduate in Meteorology from the University of Wisconsin at Madison.

Dr. Pieter Tans is a foremost world expert on greenhouse gases and CO_2 measurement. Pieter is a senior scientist in NOAA's Global Monitoring Laboratory. Since the 1970s he has worked to increase our understanding of the global carbon cycle, and he is responsible for a number of discoveries that have allowed scientists to gain more factual information about what drives the changes in our climate. In addition to his instrumental role in creating the CarbonTracker and the AirCore Atmospheric Sampling System, Pieter has overseen NOAA's monitoring of carbon dioxide in the global atmosphere from 1985 to 2019. He has worked to make the scientific evidence of rising CO_2 levels more accessible and understandable to the broader public.

Michael Banks is an author and conservationist involved in grasslands conservation primarily in Colorado and Montana. He is the author of *Malibu: Hiking along the Meaning of Life* (2004) and *Backcountry Skiing in an Era of Climate Change* (forthcoming). Michael is an avid hiker, biker and backcountry skier.

Further Readings and Videos

Beyond Debate: Answers to 50 Misconceptions on Climate Change, by Shahir Masri, Dockside Sailing Press (2018)

The Thinking Person's Guide to Climate Change: Second Edition, by Robert Henson, University of Chicago Press (2019)

Global Weirding (video series), hosted by Katharine Hayhoe: https://tinyurl.com/y33rvdfl

For current details on global climate trends, see NOAA: https://tinyurl.com/yd9qshl5

For insight into the 2020 extreme heat, see Climate Central: https://tinyurl.com/y56c5ltq

The American Meteorological Society offers many links to articles on climate: https://tinyurl.com/y5dbhjhu

See "Earth: An Operators Manual" for an excellent explanation of carbon-isotope signatures in CO2 and climate overview, beginning at 23m14s: https://tinyurl.com/yy3o2m7l

For a concise overview of the climate impact of global warming on the extinction crisis, see "WWF Living Planet Report 2020: A Deep Dive into Biodiversity in a Warming World": https://tinyurl.com/yymwqewp

For Pieter Tans's video interviews on CO2 and climate change, please see https://www.sciencehistory.org/pieter-tans

Acknowledgments

This is a small book that owes a large debt of gratitude to many who helped make the project possible. We would like to acknowledge and thank the following for their gracious assistance: the able staff at Climate Central; Dr. Jeffrey Bennett, Prof. John Harte, and Dr. Ian Bilik at Rocky Mountain Biological Labs; Prof. Maxwell Boykoff at University of Colorado–Boulder; Dr. Scott Denning at Colorado State University; Prof. Daniel P. Schrag at Harvard University. Most of all we would like to thank David Cain and Richard Larry Howe of the Citizen's Climate Lobby for their inspiring encouragement; Prof. Paul Link of Idaho State University; Christiana Kennedy at Spokes Communications; and tireless Jim Primock, whose expert publication skills brought this book to completion.

* * *

"When it is asked how much it will cost to protect the environment, one more question should be asked: how much will it cost our civilization if we do not?"

—Gaylord Nelson, founder of Earth Day

Made in the USA
Printed in Falconer, New York
March 2022